Paula Louise

from

Marie

London —
Christmas 1960

ECONOMIC PLANTS: PART I

PLANTS
THAT FEED US

BY ELSE HVASS

illustrated by
E. HAHNEWALD

BLANDFORD PRESS
LONDON

First published in the English edition 1960
© Copyright by Blandford Press Ltd,
16 West Central Street, London, W.C. 1

Translated from the Danish publication
Nytteplanter I Farver
published by Politikens Forlag 1954, and adapted for English readers
by E. B. Anderson

Printed in Holland by The Ysel Press Ltd Deventer

PREFACE

This book has been written for all those interested in economic plants and their part in the life of the world. It shows the extent and diversity of the plants necessary to supply the needs of mankind, and tells something of their origin.

A knowledge of economic plants should be an essential part of education. The editors have had this special requirement in mind in the presentation of the illustrations and in the preparation of the descriptive and botanical notes.

The colour plates are identical with those used in the Scandinavian editions, but the captions and text have been specially edited for English readers by E. B. Anderson.

The book is in two parts:

Part 1: *Plants that Feed Us* covers cereals, sugar-producing plants, vegetables, fruit and plants for feeding the animals that provide food for us, i.e. fodder plants. There is a slight overlapping with Part 2 in the last section on herbs. This has been necessary to suit the colour printing, but the reader will appreciate that in any event this section is a border case between the two volumes.

Part 2: *Plants that Serve Us* deals with plants which supply man's many needs other than food. The products are of a wide range, including medicines, dyes, perfumes, rubber, textiles and timber.

CONTENTS

PART I: PLANTS THAT FEED US

5

1 **Bread Wheat** *Triticum aestivum (vulgare)*,   1a Bearded wheat   1b Wheat grain   1c Leaf base

2 **Six-rowed Barley** *Hordeum vulgare*,   2a Two-rowed Barley *Hordeum distichon*,   2b Barley corn   2c Leaf base

3 **Rye** *Secale cereale*, 3a Rye grain 3b Leaf base

4 **Common Oat** *Avena sativa*, 4a Oat grain 4b Leaf base

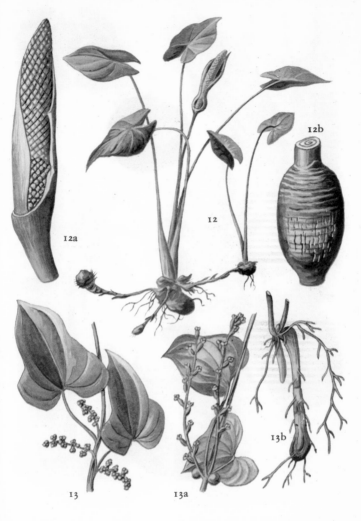

12   **Taro** *Colocasia antiquorum,* complete plant   12a Flower spadix
12b Tuber

13   **Chinese Yam** *Dioscorea batatas,* leaf and male flowers   13a Leaf and
female flowers   13b Stem with roots and young tubers

12

14a

14b

14

15

15a

16

14 **Sweet Chestnut** *Castanea sativa*, leaf with male and female flowers
14a Burr with three nuts   14b Single nut

15 **Bread-fruit** *Artocarpus incisus*, branch with male and female flowers,
together with unripe fruit   15a Ripe bread-fruit

16 **Sago Palm** *Metroxylon rumphii*, complete plant

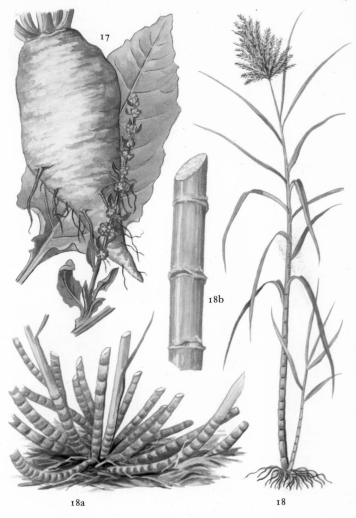

17   **Sugar Beet** *Beta vulgaris var. crassa,* flower, leaf and root

18   **Sugar Cane** *Saccharum officinarum,* complete plant in flower   18a Tuft
of sugar canes   18b Portion of stalk

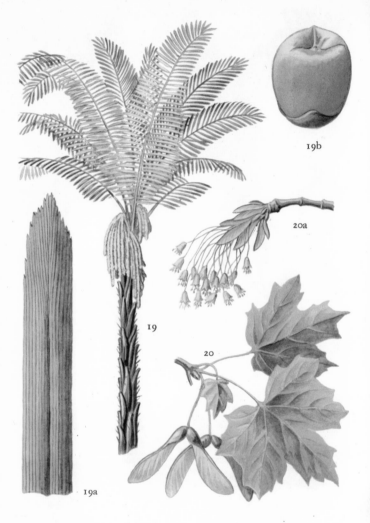

19 **Sugar Palm** *Arenga pinnata (saccharifera)*, plant with male flower
19a Apex of leaf lobe 19b Fruit

20 **Sugar Maple** *Acer saccharum*, leaf and seed (keys) 20a Flower

21   **Sunflower** *Helianthus annuus*, bud, flower and leaf   21a Seeds

22   **Rape** *Brassica napus*, flower and leaf   22a Capsule (seed pod)

23a

23b

24a   24b

24

23 **Soya Bean** *Glycine soja*, leaf and pod   23a Branch with flowers of
dark-flowered form   23b Beans of black-seeded form

24 **Ground Nut** or **Monkey Nut** *Arachis hypogaea*, complete plant
24a Nut   24b Nut with kernel

25b

25

25c

25a

25 **Coco Palm** *Cocos nucifera*, complete plant 25a Coconut with fibrous
coat 25b Nut 25c Fleshy portion of nut (copra)

26 **Oil Palm** *Elaeis guineensis*, plant   26a Portion of male flower
26b Fruits   26c Single fruit   26d Stone   26e Seed

19

27 **Olive** *Olea europaea*, flower and leaf 27a Branch with olives
27b Olive cut open

28 **Almond** *Prunus communis (amygdalus)*, flower and leaf 28a Branch
with fruit 28b Stone 28c Seed (almond)

29 **Hazel** or **Cobnut** *Corylus avellana,* twig with male catkins and bud
with female flower 29a Branch with nuts 29b Nut 29c Kernel

30 **Walnut** *Juglans regia,* branch with male catkins and female flowers.
30a Branch with fruit 30b Walnut 30c Walnut opened, showing kernel

31 **Shag-bark Hickory** *Carya ovata*, branch with male catkins   31a Nut
31b Kernel

32 **Brazil Nut (Para Nut)** *Bertholletia excelsa*, leaf and flowering branch
32a Fruit opened showing seeds (Brazil nuts)   32b Brazil nut
32c Transverse section of nut   32d Kernel

33 **Carrot** *Daucus carota*, flower and leaf 33a First year's root

34 **Parsnip** *Peucadenum sativum (Pastinacea sativa)*, flower and leaf
34a First year's root

35 **Celeriac (Turnip-rooted Celery)** *Apium graveolens var. rapaceum*
35a Celery (blanched)

36

37

36 **Swede** *Brassica napus var. napobrassicae*

37 **Beetroot** *Beta vulgaris var. crassa*

38 **Radish (Round Black Spanish)** *Raphanus sativus var. major*

38a Flower

39 **Radish** *Raphanus sativus var. radicula*, various kinds, on the left 'White Icicle'

40 **Scorzonera** or **False Salsify** *Scorzonera hispanica*, flower and leaf
40a Root

41 **Jerusalem Artichoke** *Helianthus tuberosus*, flower and leaf
41a Tubers

42   **Garden Cabbage** *Brassica oleracea var. capitata*, flower and leaf

43   **Red Cabbage** *Brassica oleracea var. capitata f. rubra*, cut open

44   **Garden Cabbage** *Brassica oleracea var. capitata*

45   **Savoy Cabbage** *Brassica oleracea bullata sabauda*
46   **Curly Kale** *Brassica oleracea var. acephala*

47   **Brussels Sprouts** *Brassica oleracea bullata gemmifera*

48   **Kohl-rabi** *Brassica oleracea caulorapa*

49   **Cauliflower** *Brassica oleracea botrytis cauliflora*

50 **Spinach** *Spinacia oleracea,* young plant   50a Female plant   50b Female
flower   50c Male plant   50d Male flower

51 **Spinach Beet** *Beta vulgaris var. cicla,* complete plant   51a Small portion
of stalk with leaf and flowers   51b Flower

52   **Cabbage Lettuce** *Lactuca sativa*

53   **Chicory** *Cichorium intybus*, flower and leaf   53a Root

54   **Chicory (Whitloof de Brussels)** *Cichorium intybus var. foliosum*

55   **Endive** *Cichorium endivis*

56   **Rhubarb** *Rheum rhaponticum*, complete plant   56a Male flower   56b
Hermaphrodite (bisexual) flower

57 **French Bean** or **Dwarf Bean** *Phaseolus vulgaris*, flower, leaf and bean
57a Green beans 57b Ripe beans

58 **Broad Bean** *Vicia faba*, flower and leaf 58a Pod

59 **Scarlet Runner** or **Runner Bean** *Phaseolus multiflorus*, flower and leaf

60   **Garden Pea** *Pisum sativum,*   60a Peas in a pod   60b Dried pea

61   **Sugar Pea** *Pisum sativum var. saccharatum,* pod

62   **Field Pea** *Pisum arvense,* flower, leaf, pod, tendrils   62a Pea

63   **Lentil** *Lens esculenta,* flower, leaf, pod, tendrils   63a Pod and seed

64a

64b

64

64   **Globe Artichoke** *Cynara scolymus*, flowering plant   64a Flower head
ready for eating   64b Flower cut open

65 **Asparagus** *Asparagus officinalis*, complete plant 65a Flowering
branch 65b Flower 65c Branch with berries

66 **Common Mushroom** *Psalliota campestris*

67b   67c

67

68b

67

68a

68

67a

67   **Red Skinned Onion** *Allium cepa*   67a White Spanish Onion
                             67b   Flower   67c Swollen stem

68   **Garlic** *Allium sativum*   68a Clove of garlic   68b Flower with sheath-
                                                 ing leaf or spathe

37

69 **Leek** *Allium porrum*

70 **Chives** *Allium schoenoprasum,* complete plant, with flower

71 **Cucumber** *Cucumis sativus,* branch with female flower and male
flowers   71a Fruit (cucumber)

72 **Vegetable Marrow, Pumpkin** or **Squash,** *Cucurbita pepo* piece of
stem with leaf and branched tendrils   72a Fruit

73   **Tomato** *Lycopersicum esculentum,* flowering branch with unripe fruit
73a Fruit (tomato)

74   **Aubergine** *Solanum melongena ovigerum*

75   **Red Pepper, Capsicum or Chilli,** *Capsicum anuum,* flowering branch
with fruit   75a Ripe fruit   75b Unripe fruit

76 **Apple** *Malus pumila,* blossom   76a, b and c Some varieties of fruit
76d Cox's Orange Pippin

77 **Pear** *Pyrus communis*, blossom   77a and b Some varieties of fruit
77c Conference   77d Louise Bonne

78 **Quince** *Cydonia oblonga var. piriformis*, blossom   78a Fruit
79 **Medlar** *Mespilus germanica*, blossom   79a Fruit

80   **Plum** *Prunus domestica*, blossom 80a River's Early 80b Victoria
                                                       80c Stone

81   **Greengage 'Reine Claude'** *Prunus domestica*, 81a Yellow Egg Plum
                                                           81b Stone

44

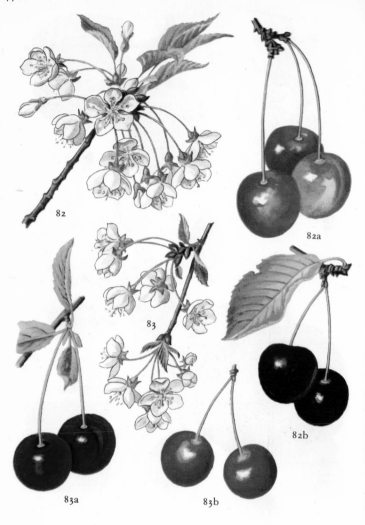

82 **Sweet Cherry** *Prunus avium*, blossom 82a Yellow and red Bigarreau
or firm fleshed 82b Black Bigarreau

83 **Sour Cherry** *Prunus cerasus*, blossom 83a Morello (black)
83b Amarelles (red)

84 **Apricot** *Prunus armeniaca*, blossom   84a Fruit   84b Stone
85 **Peach** *Prunus persica*, blossom   85a Fruit   85b Stone

86 **Black Mulberry** *Morus nigra*, branch with female flowers
  86a Branch with male flower   86b Unripe mulberry
  86c Nearly ripe mulberry

87 **Fig** *Ficus carica*, branch with fruit   87a Fig cut open (and gall fly)
                                                87b Dried fig

88

88a

88b

89

88 **Sweet Orange** *Citrus sinensis*, blossom   88a Fruit   88b Fruit cut open
89 **Mandarin** *Citrus nobilis deliciosa*

90   **Grapefruit** *Citrus paradisi*

91   **Lemon** *Citrus limonia,* 91a Flower bud and leaf   91b Fruit cut open

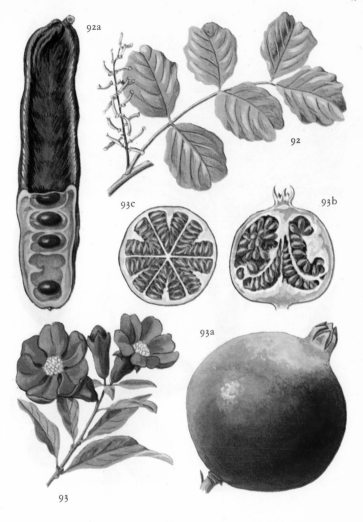

92   **Carob** or **Locust Bean** *Ceratonia siliqua*, branch with young beans
92a Ripe beans in pod

93   **Pomegranate** *Punica granatum*, blossom   93a Fruit   93b Longitudinal
section of fruit.   93c Transverse section of fruit

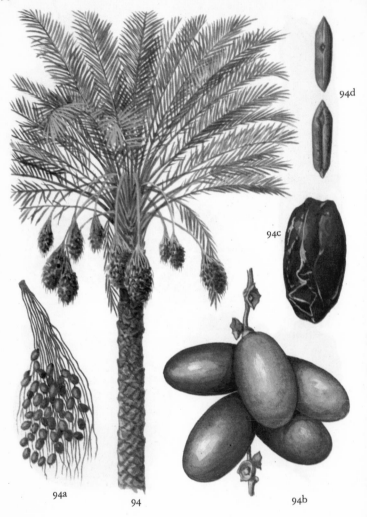

94 **Date Palm** *Phoenix dactylifera*, plant   94a Cluster of ripe fruits
94b Part of cluster of fruits   94c Dried date   94d Stones

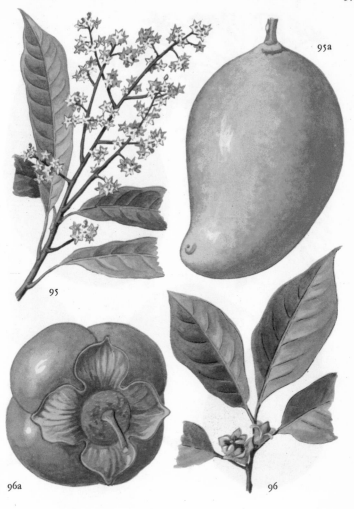

95 **Mango** *Mangifera indica,* flower and leaf   95a Fruit
96 **Persimmon** *Diospyros kaki,* flower and leaf   96a Fruit

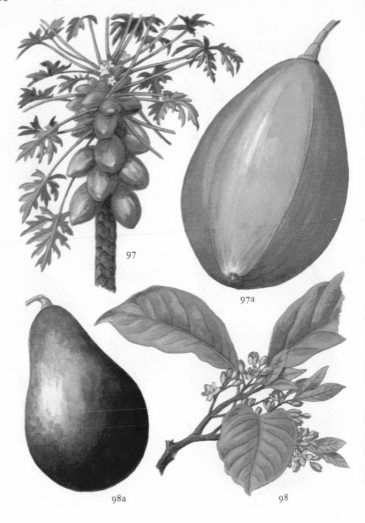

97   **Common Papaw** *Carica papaya,* flower, leaf and fruiting stem

97a   Fruit

98   **Avocado Pear** *Persea gratissima,* flower and leaf   98a Fruit

99 **Banana** *Musa sapientum,* complete plant   99a Fruit

100   **Pineapple** *Ananas sativus*, plant with fruit   100a Fruit cut open

101   **Musk Melon** *Cucumis melo var. reticulatus,*   101a Male flower
                                                         101b Female flower

102   **Water Melon** *Citrullus vulgaris,*   102a Male flower
                                                         102b Female flower

103   **Red Currant** *Ribes rubrum,* flower and leaf   103a Fruit

104   **White Currant** *Ribes rubrum var. leucocarpum,* fruit

105   **Black Currant** *Ribes nigrum,* fruit

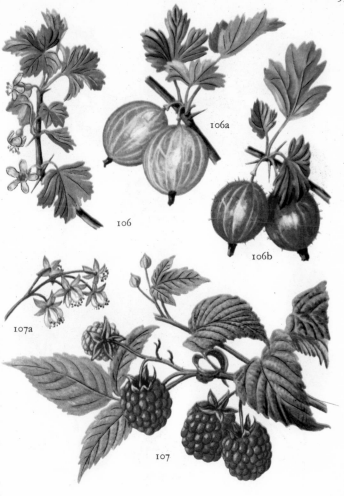

106 **Gooseberry** *Ribes grossularia*, flower and leaf   106a Green, smooth
fruit   106b Red, hairy fruit

107 **Garden Raspberry** *Rubus idaeus*, fruit and leaf   107a flowers

108   **Blackberry** *Rubus fruticosus*, flower and leaf   108a Fruit

109   **Dewberry** *Rubus caesius*, fruit

110   **Arctic Raspberry** *Rubus arcticus*, flower and leaf   110a Fruit

111   **Cloudberry** *Rubus chamaemorus*, female flower and leaf   111a Male flower   111b Fruit

112 **Wild Strawberry** *Fragaria vesca*, complete plant 112a Fruit

113 **Alpine Strawberry** *Fragaria vesca var. semperflorens*, fruit

114 **Garden Strawberry** *Fragaria x ananassa*, flower and leaf
        114a and b Two varieties of strawberry

115 115a

116a 116

115 **Sloe** *Prunus spinosa*, blossom   115a Branch with fruit
116 **Dog Rose** *Rosa canina*, blossom   116a Hips

117 **Bilberry, Whortleberry** or **Blaeberry** *Vaccinium myrtillus,* blossom
                                  117a Branch with fruit

118 **Cowberry** *Vaccinium vitis-idaea* blossom, 118a Branch with fruit

119 **Cranberry** *Vaccinium oxycoccus* blossom, 119a Branch with fruit

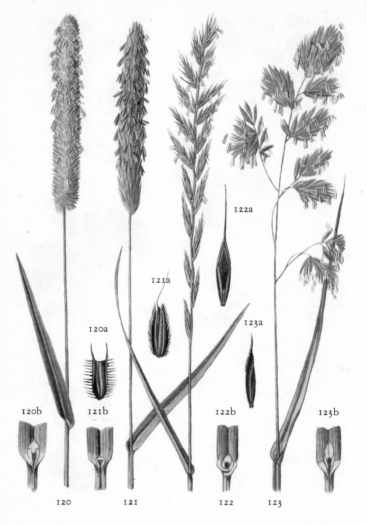

120 **Timothy Grass** or **Cat's-tail** *Phleum pratense,* 120a Seed
120b Leaf base

121 **Meadow** or **Common Fox-tail Grass** *Alopecurus pratensis,*
121a Seed 121b Leaf base

122 **Italian Rye-grass** *Lolium multiflorum,* 122a Fruit 122b Leaf base

123 **Cock's-foot** *Dactylis glomerata,* 123a Fruit 123b Leaf base

124   **Common Clover** *Trifolium pratense*, flower and leaf

125   **Alsike Clover** or **Shamrock** *Trifolium hybridum*, flower and leaf

126   **White Clover** or **Dutch Clover** *Trifolium repens*, complete plant

127 **Common Vetch** or **Tare** *Vicia sativa*, 127a Pods

128 **Lucerne** or **Alfala** *Medicago sativa*, 128a Pod

129 **Yellow Lupin** *Lupinus luteus*, 129a Pod 129b Root with bacterial
nodules

130 **Sainfoin** *Onobrychis viciifolia*, 130a Seed

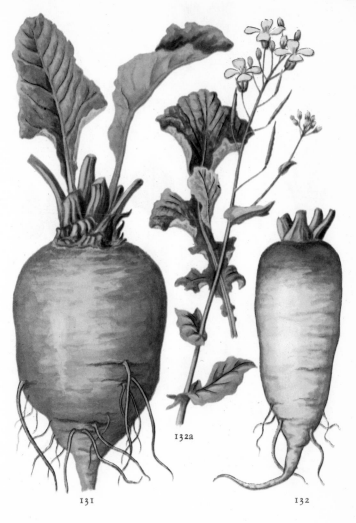

131   **Fodder Beet** *Beta vulgaris ssp. rapa*

132   **Turnip** *Brassica rapa var. rapifera,*   132a Flower and leaf

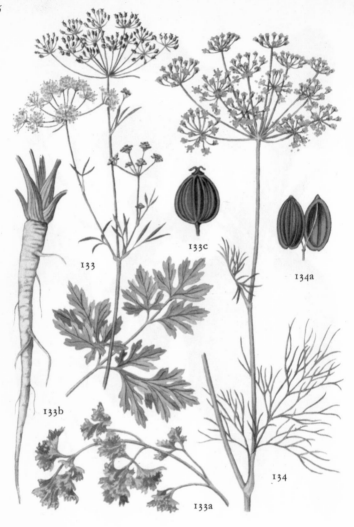

133 **Parsley** *Petroselinum hortense or crispum*, flower and leaf 133a Curled
parsley 133b Root 133c Seed

134 **Dill** *Anethum graveolens*, flower and leaf 134a seed

135 **Caraway** *Carum carvi*, flower and leaf 135a seed
136 **Fennal** *Foeniculum vulgare*, flower and leaf 136a seed
137 **Finnochio** or **Florence Fennel** *Foeniculum vulgare var. dulce*

138 **Aniseed** *Pimpinella anisum,* flower and leaf   138a Seed
139 **Chervil** *Anthriscus cerefolium,* flower and leaf   139a seed

DESCRIPTIONS OF PLANTS

The following notes are intended to supplement the information that can be obtained from the coloured plates. They deal fully with the products from the economic plants and their uses, and give information regarding geographical distribution. The botanical details have been limited to those that are of the most importance.

CEREALS

1 Bread Wheat
Triticum aestivum (vulgare)

A species of grain, close relation of one of the oldest cultivated plants, Emmer or Tokorn *Triticum dicoccum*, which was cultivated in Egypt and Iraq 5,000 years ago; it came to England about 2,000 B.C. There are many kinds of wheat: white-, yellow-, or red-seeded, awnless or with long awn (bearded wheat), spring wheat or winter wheat. New kinds are constantly being produced. Wheat must have rain in the growing season, but it needs dry weather in the ripening and harvesting season; it is never cultivated in the tropics since it is attacked by diseases in humid heat. Apart from this, the cultivation of wheat is spread over the whole world, right up to 60° north latitude, and every month throughout the year harvesting of wheat takes place somewhere. Winter wheat can give a bigger yield than spring wheat.

We speak of soft, i.e. mealy wheat, which includes winter wheat varieties of *Triticum aestivum (vulgare)*, and of hard or glassy, i.e. more albuminous wheat, which is partly found among winter and more frequently among spring varieties of bread wheat, as well as the true hard wheat, mostly grown in the Mediterranean countries.

The world's most important wheat area lies in the north temperate zone, especially in the grass steppe regions: Canada, U.S.A., U.S.S.R. and China. Three-quarters of all wheat-growing takes place in this zone. In the more northerly regions, a spring wheat is cultivated, because the winters are very cold. A spring wheat is also cultivated in the sub-tropical grass steppes (Argentina and Australia), and in sub-tropical areas of Southern Europe which have winter rain, where it is sown in winter and ripens in April to May. In Southern China, Northern India and Egypt, near the boundaries of the tropics, it is also sown in winter.

In England it is usual to grow a winter wheat, as this gives a big yield if grown, as it should be, on good and well-manured soil.

2 Six-rowed Barley
Hordeum vulgare

A long-established species of grain, which originated in Abyssinia and has been cultivated in Egypt for 6,000 years. It has practically the same distribution as wheat, yet can be cultivated farther towards the north than any other species of grain. Six-rowed barley does not give as big a yield as two-rowed barley, which originated in Asia and is a more modern variety.

Barley is sown in spring and is used chiefly for fodder. In this country it is also much used as malt in brewing beer. The barley which is to be used for malt must be two-rowed, and the grains must be large, plump, whitish-yellow with finely wrinkled surface, thin-skinned, and have a mealy kernel and good germinating power. Two-rowed barley is the commonest type of corn grown in Europe, Morocco and Abyssinia.

3 Rye *Secale cereale*

A species of grain which originated in Asia; more recently established as a cultivated plant, it possibly first appeared as a weed in the wheat fields of Western Asia. It is cultivated chiefly on the sandy glacial soils in Central Europe and in the U.S.S.R., and these two areas grow one-half of the world's production. In Northern Germany, Denmark and Poland it is the main type of corn. Rye is used not only for rye bread but also for bread made of finely sifted rye and wheat, and for crispbread. The husks are used for bran.

4 Common Oat *Avena sativa*

A species of grain which appeared possibly as a weed in the wheat fields of Western Asia, and has been cultivated in Mediterranean countries for 2,000 years. Oats bear the grain in a tuft or crest in contrast with the three types of grain previously mentioned, which carry it in ears, and they contain more fat than the others. Oats tolerate rain at the end of summer better than all other kinds of grain, and are therefore the main type grown in Norway, Sweden, Finland, the eastern part of Canada and the north-eastern part of the U.S.A. They can be cultivated on many different soils, even on bog earth. Oats are mainly used as fodder for horses and other cattle, but are also extensively used as a food when manufactured in the form of oatmeal or breakfast (rolled) oats (porridge).

5 Maize, Indian Corn or Mealies
Zea mays

A 3-15 ft tall American species of grain which before the Spaniards came to America was well-known there in numerous varieties, but probably originated from Southern Asia. It has male flowers in the upper leaves and female flower spikes (cobs) in the axils of the lower leaves. It requires a hot and humid climate and does not tolerate any frost in the growing period. Maize is cultivated much in the north temperate and sub-tropical regions, where summers are long, and in tropical agriculture by the natives. It is used principally as a fodder plant and for manufacturing purposes, but also directly for human consumption as maize porridge, maizena, corn flakes, and the unripe cobs as a vegetable. Maize is known only in its cultivated state. In all about 300 varieties are cultivated; the one illustrated is flint maize. U.S.A. is the biggest grower, but Argentine is the biggest exporter.

6 Millet *Panicum miliaceum*

A ½ ft-3 ft tall species of grain originating in sub-tropical Africa and Asia, and one of the oldest cultivated plants. In China it has been cultivated for nearly 5,000 years. It tolerates prolonged drought and is cultivated especially in the rain-deficient tropical regions of Africa and Asia. It goes into birdseed along with other species of millet, e.g. German millet.

7 Sorghum, Durra or Indian Millet
Sorghum vulgare var. durra

A 3-15 ft tall species of grain, Africa's most important bread plant. In Southern Europe and America a variety is cultivated for the manufacture of rice foods; in Southern Germany a variety is grown for green fodder. In Africa people have from ancient times made beer from it and in China paper.

8 Rice, *Oryza sativa*

A 1½-5 ft tall species of grain,

probably a native of India. It has been cultivated in China for nearly 5,000 years. Rice is an annual, tropical swamp plant, but can be cultivated in sub-tropical regions, if summers are sufficiently warm. It is germinated and the young plants, when large enough, are transplanted into flooded fields, which are drained at harvest.

After wheat, rice is the world's most important species of grain, and it provides food for more people than any other plant in the world.

Rice grains contain much starch and are used for the manufacture of rice flour. Rice is husked, but where it constitutes the chief part of the diet it ought not to be polished since B-vitamins are thereby removed, and the deficiency disease, beri-beri, would appear among any population depending mainly on it for food. Rice straw is used for cigarette paper.

Rice is cultivated especially in Southern and Eastern Asia. In many places it is harvested two to four times a year; but that impoverishes the soil. Only a small part of the world's harvest comes on to the market. China and India cultivate most, but Burma and Siam export most. The rice illustrated is a short-grained, Chinese kind.

9 Potato *Solanum tuberosum*

A 20 in. tall vegetable with stem tubers. Originally a native of, and cultivated in, mountainous regions of the Andes in Peru and Chile, it was exported to Europe after the discovery of America, and is now widely distributed to about the same extent as rye. Potatoes contain about 17½% starch and a certain amount of vitamin C. They are one of the most important foodstuffs in Europe, where they first acquired proper importance after the famine of 1770. Potatoes do not play a very great part in world trade, since they cannot

be stored for a very long time; North America, Holland, the Canary Islands, and Mediterranean countries, however, do export early potatoes. Germany, Poland and U.S.S.R. have the highest production of potatoes.

10 Sweet Potato *Ipomea batatas*

A creeping plant with root tubers growing on the runners (which are quite 3 ft long), originating from South or Central America. It is of great importance in the tropics, where it takes the place of potatoes. Sweet potatoes contain 20% starch and 5% sugar, hence their name. Flour and Brazilian arrowroot can be manufactured from them. China grows one-half of the world's production. In U.S.A. they are cultivated on a large scale in the southern states, and in Africa in the savannah regions.

11 Tapioca, Manioc *or* Cassava *Manihot utilissima*

A 6½ ft tall, perennial, tropical rain-forest plant with root tubers, originating from Brazil which plays an important part as a food plant in the tropics. The large tubers can weigh 11 lb-22 lb. They contain prussic acid, which is poisonous and must be removed by washing, or destroyed by roasting or drying, before the edible meal can be manufactured from them. The meal is called sago meal and is used for milk puddings and for thickening juices of such red fruit as raspberries and red currants.

The extracted juice of the tubers, after evaporation by boiling, can be used as a preservative for meat.

12 Taro *Colocasia antiquorum*

A perennial herb with swollen root, rich in starch, originating from Indonesia. It is cultivated chiefly on mountain slopes in moist or artifically watered tropical regions of Africa and Asia. The tubers can weigh 4½ lb-9 lb and contain 50 % starch. Taro meal or Portland arrowroot is manufactured from the tubers.

13 Chinese Yam *Dioscorea batatas*

A twining plant with root tubers, which is widespread in the tropics and is the most important tuberous plant among the Negro tribes in the Cameroons and Togoland. Yams require high temperature and humidity during their 10-months growing period. The tubers contain the poison dioscorin, which can be removed by boiling. Yams tubers are called Chinese potatoes; they are sweeter than ordinary potatoes, and of better keeping quality than sweet potatoes. The meal is called dioscorea-arrowroot, South Sea arrowroot or Guiana arrowroot.

14 Sweet Chestnut *Castanea sativa*

A 30-100 ft high tree, growing wild in Southern Europe and Asia Minor, and introduced into Central Europe and the Scandinavian countries. The fruits are nuts (chestnuts). They are found two or three together in a prickly case and contain 50 % starch and 4 % fatty oil. Chestnuts are eaten boiled or roasted, and chestnut flour can be manufactured from them. The best kinds are obtained from Italy and Southern France. Chestnut wood is used for furniture.

15 Bread-fruit *Artocarpus incisus*

A 40-55 ft high tree with milky sap, a native of Indonesia, but has since early times been cultivated in other tropical regions, especially in the islands of the Pacific Ocean. The fruits of two or three trees are sufficient to feed one man. Bread-fruits are a combination of the fleshy perianth leaves of the female flowers after flowering, with the axis of the inflorescence, so that the whole becomes a 'false fruit', about 16 in. long and 10 in. in cross section, and with a weight of 1 lb-4½ lb. Bread-fruits contain 17 % starch and are eaten roasted, but can also be preserved.

16 Sago Palm *Metroxylon rumphii*

A 26 ft-40 ft high palm, growing wild in Indonesia, where it is also cultivated, and in New Guinea, where it plays an important role as food. True sago is extracted from the stem. The palms are felled when 20-30 years old, immediately before they flower, when they are filled with starch-containing pith. A single stem can give 650 lb-900 lb of pith. A large proportion of the sago on the market, however, is not genuine sago, but manufactured from potato or tapioca flour.

SUGAR-PRODUCING   PLANTS

17   Sugar Beet
Beta vulgaris var. crassa

A biennial herb, which in the first year accumulates stored food in its taproot, of which 17-27% is sugar. Sugar beets are deeply rooted in the ground and therefore tolerate drought. Cultivation of sugar beet started in Germany 150 years ago with the cultivation of the European wild beet, but at that time the roots contained only 5-7% sugar. After the continental blockade in the 1800's, the cultivation of sugar beet in Europe advanced immensely. It was developed in Britain between 1924 and 1934 and during the Second World War the beet sugar crop was of vital importance. It grows particularly in Eastern England. The waste from sugar manufacture is used for cattle fodder. It is primarily a product of Europe, the U.S.S.R. and the United States.

18   Sugar Cane *Saccharum officinarum*

A 12-20 ft tall grass which originated in India, where it has been cultivated for 3,000 years. The stems are about 2 in. in diameter and filled with pith with a 20% sugar content. The amount of sugar is greatest immediately before flowering, and the canes are harvested at that time. Plants are propagated by cuttings. Sugar cane can only be cultivated in the tropics or the hot subtropics. The growing period is 12-18 months. It requires abundant moisture, except in the ripening period. India and Cuba grow one-half of the world's production of cane sugar, one-half of the world's exports goes from Cuba to the U.S.A. World production of cane sugar is twice as great as of beet sugar, because of the cheap labour in the tropics and consequent greater competitiveness. Beet sugar and cane sugar are chemically of the same substance.

19   Sugar Palm
Arenga pinnata (saccharifera)

A palm which is a native of Indonesia and India. The sugar is extracted by cutting off the male flower stalks at their bases and collecting the sap which amounts to about 7 pints a day. Palm sugar can be extracted from the sap by evaporation or it can be fermented to palm wine (toddy).

20   Sugar Maple *Acer saccharum*

A forest tree growing wild in Canada and the eastern part of the U.S.A. The trees must be 50 years old before they can stand the yearly tapping, which takes place in February-March by drilling into the bark at a height of about 3 ft. Each tree can give 4-6 lbs of maple sugar, which tastes and smells of coumarin, a crystalline substance.

21   Sunflower *Helianthus annuus*

A 3-10 ft tall, annual plant with composite flowers, a native of Mexico and Peru, often cultivated as an ornamental plant in this country. It was brought to Europe by Spaniards and is cultivated especially in Southern Russia, Hungary and Italy. The composite flower heads can be up to 14 in. in diameter, the fruits, 'sunflower seeds', are botanically nuts, the seeds of which contain 40-50% fatty oil. Varieties producing edible seeds and those suitable for oil extraction are cultivated. The oil extracted by pressure is used as an edible oil in Russia and as an oil substitute in some countries. The cakes which remain after oil extraction are used as cattle fodder.

22 Rape
Brassica napus

An annual or biennial herb, a close relation of the swede, which originated in North-west Europe. It is cultivated especially in India, eastern Asia and Denmark. The seeds contain 35-40 % fatty oil, which is extracted by pressure. Rapeseed oil was used in the past for lighting and is now used as an edible oil, for margarine and lubrication. Rape must be harvested when it is ripe, but not over-ripe, since in that way the seeds are lost; the harvesting should take place early in the morning. Rapeseed cakes remaining after the oil has been expressed have great importance as fodder.

23 Soya Bean
Glycine soja

The plant may be from 9 in.-2 ft high, but most kinds are less than 1½ ft. It is an annual leguminous plant, a native of eastern Asia, and has been cultivated in China for 5,000 years. Soya beans are a common food in China and Japan. In 1880 the plant was introduced into North America, where it is used as green fodder. The flowers are whitish or light violet. Soya beans can be yellow, brown or black. They contain 30-40 % protein and 12-24 % oil, which is expressed, or extracted with benzine, and used for margarine or soap. The residue (cake) after pressing is used for fodder. Soya oil is a semi-drying oil and can therefore be used in the paint, lacquer and varnish industries. In China and Japan a milk substitute is made from the crushed beans, and shoyu sauce or soya from boiled, split, and fermented beans blended with toasted wheat flour; the latter product is imported into Europe.

24 Ground Nut or Monkey Nut
Arachis hypogaea

1-2½ ft high, annual leguminous plant, which probably originated in Brazil. Ground nut plants set seed within four months, and before ripening the pods thrust themselves down into the ground; they are ¾ in. to 1½ in. long and are called peanuts. The seeds contain 40-60 % oil which by cold pressing is extracted as a fine, light oil suitable for edible purposes and margarine. By using greater pressure and heat a darker oil is obtained suitable only for soap manufacture. The African peanuts give the best oil, the Japanese give the biggest yield. The oil cakes are a valuable cattle fodder. It is cultivated in the tropical savannah regions in Africa, in the lowlands of Northern China and in U.S.A.'s cotton belt in the southern states.

25 Coco Palm *Cocos nucifera*

A high, slender palm (60-75 ft), which originated probably from Colombia, but has been cultivated in south-east Asia for 3,000-4,000 years. It is an oil plant of world-wide importance and widespread in the equatorial tropics. It grows in coastal regions and tolerates salt water. Coconuts hang in clusters and are stone-fruits. The fibrous coat of the shell is used for brushes, mats, brooms, etc. Coco palms bear fruits when 7-70 years old and these ripen 10-12 months after flowering. Copra is the dried white flesh of the seed. It contains 63 % fatty oil, which is expressed at about 125°f. In the tropics coconut oil is used for food, lamp oil and as a cosmetic; in Europe it is used in the manufacture of margarine, palmitin, soap and stearin candles. 'Desiccated coconut' is used in domestic cookery and in the chocolate industry. Formerly the producing countries exported only the copra, but now they themselves manufacture oil for export (especially Malaya and Ceylon). The Philippines are the biggest exporting country and sell more oil than copra. The young top shoot of the coco palm is eaten as 'palm cabbage' and it tastes like nut kernels.

26 Oil Palm *Elaeis guineensis*

A high, slender palm (75-app. 100ft) growing wild in tropical Africa, especially in the coastal regions. The fruiting branch bears 800-4,000 plum-like stoney fruits. Both the fruit pulp and the seeds themselves, the palm kernels, are oil-bearing, but the palm kernels give the finest oil, which smells of violets. Fresh palm oil is yellow or orange-red and is used for food, margarine, soap, candles and cosmetics. Oil palms are now cultivated in Brazil, Malaya and Indonesia, whence palm oil is exported. Both palm oil and palm kernels are exported from the Guinea Coast.

27 Olive *Olea europaea*

A tree up to 30 ft high, growing wild in the sub-tropical Mediterranean regions, where it is cultivated and water is reasonably plentiful. It is pruned to a height suitable for picking the fruit. The olive tree can live 2,000 years and is a very ancient cultivated plant. The fruit is a stone-fruit the size of a small plum, and both the flesh of the fruit and the kernel of the seed contain oil, which is extracted by pressing. The first pressing (cold pressing) gives the finest oil, which is clear and colourless and is used as an edible oil. Warm pressing gives an inferior edible oil used also for lubricating oil and for soap manufacture. Olive fruits are also eaten whole in the pickled state, both green unripened and black fully ripened. Olives are now also cultivated in the sub-tropical regions of America, especially California.

28 Almond
Prunus communis (amygdalus)

A tree of medium height, growing wild in Afghanistan, Turkestan and Iran, and commonly found in the Southern European countries, California and in the oases of Western and Central Asia. The fruit is a stone-fruit with leathery flesh and hard stone. The stone has a thin shell in the variety *fragilis,* the brittle-shelled almond. Sweet almonds are the seeds of *Prunus amygdalus var. dulcis;* they contain 50% fatty oil and are used in domestic cookery and for marzipan. From them almond oil can be pressed out, the residue after pressing is almond bran, and both are used for cosmetics. Bitter almonds are obtained from the variety *amara,* which contains prussic acid and bitter almond oil.

29 Hazel or Cobnut *Corylus avellana*

A bush up to 12 ft high, widespread in Europe and Asia Minor, and often growing wild in our woods. The fruit is a nut and the seed contains 60% fatty oil. Besides being used for eating, the nut kernels are useful for the oil which can be extracted and utilized for oil colours and in perfumery.

30 Walnut *Juglans regia*

A 30-100 ft high tree, a native of Asia and Northern Greece. The fruits are stone-fruits the size of plums, the walnuts are the stones. In fresh walnuts there is 50% of fatty oil and in the dried 65%, which can be expressed and is used as an edible oil and for fine oil colours. The most important producers of walnuts are Southern France and Yugoslavia; but the tree is cultivated in the whole of Europe, among other reasons on account of its wood.

31 Shag-bark Hickory
Carya ovata

A North American tree of the same family as the walnut. The fruits resemble rather long walnuts and many species are cultivated for the sake of their nuts, especially the thin-shelled, which is the one illustrated. From the seeds oil can be extracted — the American walnut oil which is used in the soap industry. The tree is also cultivated on account of its wood.

32 Brazil Nut (Para Nut)
Bertholletia excelsa

A huge, tropical tree with buttress-like roots, growing wild in Brazil and Venezuela, especially alongside rivers. The fruit is spherical, woody, and the size of a small child's head, and the para nuts or brazil nuts, are the seeds, which are triangular and hard-shelled and lie closely pressed together within the fruit. The kernels, which we eat, are the white flesh of the seed, which contains 70 % oil.

VEGETABLES

33 Carrot *Daucus carota*

A 12-40 in. high, biennial umbelliferous plant, a native of the temperate regions of Europe and Asia, and long establish-ed in cultivation. The thick root develops the first year. The yellow colour is due to the carotene, which when eaten forms vitamin A. The nutritive value of carrots is 180 cal. per lb. Carrots are almost conical in shape. They can be eaten cooked as a vegetable or raw in salad. There are numerous kinds. The large coarse, white or green carrots are used as fodder for cattle.

34 Parsnip
Peucadenum sativum (Pastinacea sativa)

A biennial umbelliferous plant, com-monly growing wild in Southern and Central Europe. The first year's tap root contains cane sugar and glucose and a little fatty oil, together with essential oil, and can be used for the manufacture of coffee substitute, but is mainly used for culinary purposes. Up to the eighteenth century it was much cultivated, but was supplanted by the carrot, in spite of the fact that the pars-nip is a much more frost-resistant plant.

35 Celeriac (Turnip-rooted Celery)
Apium graveolens var. rapaceum

A biennial umbelliferous plant, grow-ing wild in meadows near the sea in Europe. The hypocotyl, the upper part of the root, together with the lower part of the stem, is swollen to form a large tuber rich in food value, and with a characteristic flavour due to essential oils. Celery is a plant long established in cultivation, and was already well-known in ancient Egypt. Blanched celery *Apium graveolens var. dulce* is the

original cultivated form of celery. White leaf-stalks are produced by covering the plant with earth or straw, so that the flavour becomes mild and the consistency crisp. These are eaten raw, or cooked as a vegetable.

36 Swede
Brassica napus var. napobrassicae

A biennial whose cultivation has not been established very long. It is rapid-growing, a so-called 3 months' plant, and easily satisfied, as it can grow in sand, humus or peaty ground. It is used as fodder for fattening cattle. It is very rich in vitamin C. The swede is much cultivated in England and Denmark, where it is gradually sup-planting the turnip.

37 Beetroot
Beta vulgaris var. crassa

A biennial plant, which is related to sugar beet and fodder beet. In its first year it forms a root tuber, the red colour of which is due to red cell sap. Beetroot has, like the other beet, false annual rings of many layers of cambium round one another. It has no great nutritive value, but is cultivated throughout Europe and used in salads.

38 Radish (Round Black Spanish)
Raphanus sativus var. major

A biennial cruciferous plant developing late in the year, probably a native of Asia. It is a plant long established in cultivation, and was depicted in the pyramid of Cheops 2,700 years before our era. There are red, white and black kinds. The flavour is due to its mustard oil content, and the part eaten is the thickened root. Radishes have no special nutritive value, but they are popular and eaten in salads.

39 Radish
Raphanus sativus var. radicula

An annual cruciferous plant developing early in the year. It probably originated from the same species as the Round Black Spanish radish and first appeared on the north-west coast of Europe in the sixteenth century. The 'radish' consists mainly of the thickened hypocotyl; its nutritive value is small, but its vitamin C content can be fairly large. They are used in salads.

40 Scorzonera or False Salsify
Scorzonera hispanica

A perennial plant with composite flowers and milky sap, a native of Central and Southern Europe. In the first year the plant forms the edible taproot, which is cylindrical and 12 in. long, with a black corky layer outside. The leaves can take the place of mulberry leaves as food for silkworms. Scorzonera is a comparatively new cultivated plant and is only grown very little in this country.

41 Jerusalem Artichoke
Helianthus tuberosus

A perennial plant with composite flowers, a kind of sunflower. It is a native of North America, and came to Europe after the Thirty Years War. It practically never blooms. It has short subterranean runners, at the tips of which are potato-like stem tubers, sweetish and fairly nutritious, and not damaged by frost. The plant is less exacting in its requirements than the potato as regards climate and soil. In Belgium, Jerusalem artichokes are used for the manufacture of alcohol. The example illustrated is without runners.

42 Garden Cabbage
Brassica oleracea var. capitata

A biennial, cruciferous plant, which is one of the most nutritious of vegetables. The cultivation of cabbage was already widespread in Mediterranean countries in ancient times. All cultivated forms are biennials, and in the first year they accumulate stores of food in their leaves, forming at maturity a close heart, the shape depending on the variety.

43 Red Cabbage
Brassica oleracea var. capitata f. rubra

A pickling cabbage, with food stored in its leaves. Red cabbage differs from garden cabbage practically only in its taste and in the red colour of its cell-sap. The colour changes with the degree of acidity; if red cabbage is boiled in ordinary water, the colour becomes blue, but if vinegar is added to the water it becomes red.

44 Garden Cabbage
Brassica oleracea var. capitata

A variety of cabbage known as 'drumhead', which forms a very compact head. The nutritive value is 135 cal. per lb. Cabbage is one of our most important vegetables for cooking. German Sauerkraut is prepared from a drumhead type of cabbage.

45 Savoy Cabbage
Brassica oleracea bullata sabauda

This bears a resemblance to drumhead cabbage, but has a less compact head. It is hardier and easy to cultivate. It has been well-known since the Middle Ages.

46 Curly Kale
Brassica oleracea var. acephala

A leafy cabbage 4-32 in. high, which resembles most closely the original source of all garden cabbage and is a very old-established vegetable for cooking. It is hardy and tastes best when it has had frost on it; it is a valuable source of vitamins A and C in winter.

47 Brussels Sprouts
Brassica oleracea bullata gemmifera

A species of cabbage 39-40 in. tall, which no doubt originated in Belgium. It bears 20-80 lateral buds on its stem, which grow into small spherical heads (sprouts). When these are pulled off, the plant develops afresh, and several crops of sprouts can be gathered. The plant tolerates frost. The nutritive value of Brussels sprouts is 315 cal. per lb.

48 Kohl-rabi
Brassica oleracea caulorapa

This stores food in a tuber-like, swollen stem, above the ground, which can be spherical or oblong in shape; it bears numerous leaf-scars. The colour can be green, white or violet. Kohl-rabi is very hardy, but is cultivated very little in this country, and practically only as a fodder plant. In Germany it has great importance as a human food.

49 Cauliflower
Brassica oleracea botrytis cauliflora

A species of cabbage, which is peculiar as the inflorescence grows in the plant's first year and forms the cauliflower head, in which the majority of the branches are full of nourishment. Its nutritive value is 157.5 cal. per lb. Its age as a cultivated plant is doubtful; some believe that it was cultivated in ancient times by the Greeks, and that it must have originated in Asia Minor; others assert that it first appeared in the early 16th century in Cyprus.

50 Spinach *Spinacia oleracea*

An annual plant, comparatively recently cultivated in Europe, which probably originated in Western Asia, whence it spread in the 1500's. The female plants are the most valuable, as they develop more slowly than the male plants and have bigger leaves, which form a rosette and do not wither until the seed ripens. Spinach leaves are rich in vitamins A and C, and besides being used as a vegetable can also be used for the manufacture of vitamin preparations. Spinach contains a certain amount of oxalic acid and iron.

51 Spinach Beet
Beta vulgaris var. cicla

A very long-established cultivated plant, well-known in Mediterranean countries from early times; its cultivaiton in the north (Scandinavia) is also long-established. Spinach beet is a close relation of sugar beet, fodder beet and beetroot; but it is the leaves which are used — as salad or spinach (Roman spinach) It produces good leaves in middle or late summer, when ordinary spinach is over.

52 Cabbage Lettuce *Lactuca sativa*

A plant with composite flowers and a little milky sap. It originated possibly in Southern Europe; and is a very old-established cultivated plant. There are many kinds, some entirely without heart, others with more or less firm hearts; in this country cabbage lettuce with heart is chiefly cultivated. Its nutritive value is very small, but the vitamin C and carotene content is large.

53 Chicory *Cichorium intybus*

A wild perennial plant with composite flowers and milky sap. In its cultivated state it is grown as a biennial. The root contains in its first year a certain amount of starch and sugar, which on roasting is changed into a caramel-like substance; it has, therefore, been used for manufacture of a coffee substitute since 1722. It is also used in a mixture with ground coffee. It is cultivated in South Jutland, Holland, Belgium and Germany.

54 Chicory (Whitloof de Brussels)
Cichorium intybus var. foliosum

This originated from the wild chicory and is a variety of No. 53. The bleached elongated head of crisp leaves develops in the winter at a temperature of about 50° f. under a covering of sand and earth, or in a cellar or dark shed, which makes it crisper and less bitter than the green shoot. It is used as salad. This winter salad is cultivated only a little in this country, but it is imported from Belgium and France.

55 Endive *Cichorium endivis*

A cultivated plant, well-known to the ancient Romans and Greeks, which probably originated in Egypt. There are various kinds. The illustration is of curled endive, *var. crispa,* which can be used like cabbage lettuce, although it has a more bitter taste. In this country it is blanched for use. Large quantities are grown in Western Europe, e.g. Holland and France.

56 Rhubarb
Rheum rhaponticum

A large perennial herb, which originated in Central Asia and first came into cultivation in Europe as a food plant in the middle of last century. The leaf stalks (which contain oxalic acid) are used in fruit pies or stewed, or for bottling and jam. The kind illustrated is the somewhat coarse Victoria, but there are several finer, better flavoured, kinds.

57 French Bean or Dwarf Bean
Phaseolus vulgaris

An annual cultivated plant, a native of tropical and sub-tropical America. It was cultivated in the empire of the Incas in Peru, and first came to Europe in the 1500's. There are two distinct types, tall-climbing and dwarf. The pods are eaten in the unripe state as a vegetable.

There are also a yellow-podded, or wax pod, and a green-podded snap bean. With certain dwarf varieties the dried ripe seeds are used; they are known as white haricots and brown haricots, and have the same high nutritive food value as dried split peas. Cultivation is widespread in Europe.

58 Broad Bean *Vicia faba*

An annual plant, about 3¼ ft high, growing wild in Asia and Africa, no doubt introduced from the regions south of the Caspian Sea. A very old-established cultivated plant in Europe. It was introduced into Britain many years ago and is referred to by Chaucer. The broad bean has large, light-coloured seeds which are used for human food.

59 Scarlet Runner or Runner Bean
Phaseolus multiflorus

A tall, climbing plant, which originated from tropical America and is less hardy than the French bean. Both are susceptible to frost, but the runner bean should be sown a little later, from May onwards. The pods form excellent beans for slicing.

60 Garden Pea
Pisum sativum

A leguminous plant growing up to 6½ ft high. It is a very old-established cultivated plant, which was well-known in ancient Egypt, and possibly originated from southern Europe. The seeds (peas) contain much starch and protein and are picked unripe for cooking, being succulent, sweet and very nutritious. There are many different kinds.

61 Sugar Pea
Pisum sativum var. saccharatum

A type of garden pea without the tough membrane inside the pod, so that the pod presses against the seeds gradually as they grow. The whole pod with its small peas inside is eaten unripe as a vegetable, boiled, or occasionally raw in salads.

62   Field Pea, *Pisum arvense*

A leguminous plant, which sometimes grows wild in Mediterranean countries. The peas are dry and bitter and cannot easily be cooked. The whole plant is used as green fodder or hay, or for digging in to improve the soil, as it has nodules containing nitrogen bacteria on the roots.

63   Lentil *Lens esculenta*

A small, annual, climbing plant, probably a native of Asia. Its cultivation goes a long way back in time and it played a part in the history of the Jewish patriarchs. The pods are only ½-¾ in. long and contain 1-3 flat seeds, which are white, green or red-brown and very nutritious. In Roman Catholic countries lentils provide people with a palatable and nutritious food in Lent.

64   Globe Artichoke *Cynara scolymus*

A perennial plant, about 3¼ ft high, with composite flowers, which probably originated from Mediterranean countries. Shortly before the flowers expand, the heads are cut off and boiled whole. The part eaten is the lowest thick part of the leaves of the involucre or head and the base of the receptacle. The Romans grew globe artichokes, and their cultivation was revived in Italy in the 1400's. In this country they are a long-established product of market gardens.

65   Asparagus *Asparagus officinalis*

A member of the same family as Lily of the Valley; its cultivation has been long-established and it was well-known even in ancient Rome. The part eaten is the future aerial shoot, which when covered with earth becomes white, succulent and crisp. If it grows above ground, it becomes green and is eaten, boiled, as green asparagus. An asparagus bed must stand two years, before the plants can be 'cut', and then it can be cut twice daily from May to Midsummer Day. The fine flavour is due to its asparagin and sugar content. Its nutritive value is very small, (10.35 cal. per lb) but it contains B-vitamins. It is especially cultivated in North America and France.

66   Common Mushroom
Psalliota campestris

The only fungus which is cultivated on a large scale the whole year round. It is a saprophytic plant, which is cultivated on horse manure in dark cellars or sheds, where it is easy to maintain constant temperature and humidity. The part eaten is the aerial fruiting body. Formerly mushrooms were imported from France, but now we are self-supplying, both for fresh and preserved mushrooms.

67   Red-Skinned Onion
Allium cepa

A biennial cultivated plant, growing wild in Western Asia, which was well-known to the ancient Egyptians. Its flavour is due to volatile, sulphur-containing onion oil. The onion is really the basal portion of the leaves, and is full of stored food and very rich in calories. Onions are cultivated in all temperate countries. Depending on the variety, the onion may have a skin varying from bright red to pale straw.

68   Garlic *Allium sativum*

A plant long established in cultivation, which originated from the steppes of Western Asia. Garlic is the most nutritious of all culinary vegetables, with 585 cal. per lb, and contains more protein than green peas. Garlic has many subordinate bulbs or 'cloves', which are all encased in the scale leaves of the parent bulb. The strong flavoured oil in garlic gives people who eat much of it an unpleasant odour. It is cultivated especially in southern and eastern countries.

69 Leek *Allium porrum*

A mild-flavoured species of onion, which originated from Mediterranean countries and was already well-known as a culinary vegetable in ancient Rome and Egypt. It forms no typical onion bulb, but long, succulent, nutritious leaf sheaths, which are blanched by earthing up and then used as a vegetable or in soup.

70 Chives *Allium schoenoprasum*

A species of onion 6-12 in. high, which grows in tufts, and is found growing wild from Southern Sweden to Corsica and from Western Europe to Eastern Siberia. The tubular green leaves are used raw as a flavouring herb which is milder and subtler than the flavour of onions. Chives are rich in vitamins A and C.
They are easy to cultivate.

71 Cucumber *Cucumis sativus*

An annual climbing plant, which originated from the northern East Indies and was even in ancient times cultivated in Mediterranean countries. The fruit is botanically a succulent berry with numerous seeds in six rows. Its water content is 97%, and its nutritive value is very small, only 6.3 cal. per lb. In this country they may be grown outdoors, or preferably in frames or greenhouses. They are used mainly for salads. A small fruited variety provides gherkins used for pickling.

72 Vegetable Marrow, Pumpkin or Squash
Cucurbita pepo

An annual climbing plant, a native of Central America from Mexico to Peru. The fruits are botanically large berries, of various shapes and colours. The flesh is firmer and less succulent than that of the cucumber and melon. It is eaten as a vegetable, boiled or stuffed and baked, or for making marrow and ginger jam. Edible oil can be extracted from the seeds. The variety *Cucurbita pepo aurantia* is cultivated for the beauty of its fruits, which are known as ornamental gourds.

73 Tomato
Lycopersicum esculentum

An annual plant, closely related to the potato, a native of Mexico and Peru, and which came to Europe in the 1500's. It was cultivated earliest in England and Southern Europe; only at the end of last century did its cultivation spread into Central and Northern Europe. The fruit is botanically a berry. Tomatoes have no great nutritive value, but their great richness in vitamins A, B and C makes them of great importance for health. This country is a very big producer of tomatoes, which are mostly grown in greenhouses; in good summers outdoor tomatoes can be cultivated.

74 Aubergine
Solanum melongena ovigerum

An annual cultivated plant, which is related to the potato plant and originated in tropical India and the East Indies. The fruit is a large berry, which can weigh 2¼ lb, and it can vary greatly in shape and colour. In this country it is seldom cultivated or eaten, but it is very widely grown in Southern Europe and other warm countries. As a dish, the aubergine is cut up into slices and fried in oil.

75 Red Pepper, Capsicum or Chilli
Capsicum annum

An annual South American cultivated plant, which in 1585 came to Hungary, where it is still used in a large number of dishes. The ripe, swollen berry fruits of sweet capsicum (¾-8 in. long) can be green, yellow, red or violet and are used more and more in the fresh state for salads. Paprika is the dried and powdered fruits; cayenne pepper is made from the small-fruited forms known as chillies. Capsicum is very rich in vitamin C. It is much cultivated in Hungary, Mediterranean countries and in the tropics.

FRUIT

76 Apple *Malus pumila*

This fruit tree, a very old cultivated plant, was originally grown in the countries around the Mediterranean. The apple is a fruit containing seeds or pips; ripe apples contain 7-9 % of sugar, some acid, and 82-85 % of water. The nutritive value is as in cabbage, 100-320 calories per lb. Apples, especially winter apples, are of special value for their keeping quality. The most important fruit of the temperate zone, they are chiefly grown in Europe, Canada, U.S.A., Australia and South Africa. There are a great many varieties, Cox's Orange Pippin being one of our best eating apples. Apples are also used for the preparation of juice and cider.

77 Pear
Pirus communis

A fruit related to the apple; numerous varieties were grown in ancient Rome and Greece. The pear is a fruit containing seeds or pips with a poorly developed core. There are many varieties, some suitable for eating, others for cooking.

78 Quince
Cydonia oblonga var. piriformis

A small fruit tree of the seeded-fruit family, which comes from South-east Europe and South-west Asia. It is a very old cultivated plant which was much valued. The fruit may be shaped like an apple, or pear-shaped like that shown in the illustration, and has a very strong odour. Quince is grown in England only to a small extent. The fruit may be used for stewing, or for making jam, marmalade or quince-bread.

79 Medlar *Mespilus germanica*
A small tree, coming from South-east Europe and South-west Asia, now grown mainly in Central and Southern Europe. The core is developed like a stone around the ovaries, so that the medlar-fruits have 5 stones. They can be eaten only when they are over-ripe or bletted; or used for making medlar jelly or cheese.

80 Plum *Prunus domestica*

Part of the stone-fruit family. Its original habitat was probably the countries to the north and south of Caucasus; it was known in ancient Rome and in Mesopotamia. The plum is a juicy stone-fruit, which may be eaten fresh or as a preserve, or dried, in the form of prunes. There are many varieties, most valuable of which are the Pershore or Yellow Egg for jamming and bottling, and the free-cropping Victoria, good enough to eat raw. The finest dried blue plums are called French plums. Hungary, Yugoslavia, Turkey, California and South Africa are extensive plum growers and exporters of dried plums.

81 Greenage ('Reine Claude')
Prunus domestica

Greengages are plums with juicy, green or yellow fruit, and are the finest dessert plums. There are now varieties which crop more freely in this country than the one illustrated.

82 Sweet Cherry *Prunus avium*

This fruit tree is a variety of the wild bird-cherry. It comes from the East and Greece, where it has been grown since ancient times. Cherries are stone-fruits with 10 % of sugar. Yellow or red bigarreaus have firm flesh; both the varieties illustrated belong to this species, as do Spanish cherries. Heart or Gean cherries have soft flesh. Cherries are generally eaten fresh or bottled, but are also used for jam-making and the manufacture of liqueur, especially Cherry Heering, which is a world-renowned export article from Denmark.

83 Sour Cherry *Prunus cerasus*

A fruit tree originating from Asia Minor, which came to Rome in 73 B.C., and is cultivated only in Europe. Among the varieties grown there is the Morello, with black fruits; it is also the variety most widely grown here. They are usually stewed. Amarelles with red fruit are widely grown in Sweden, but seldom in England.

84 Apricot *Prunus armeniaca*

A small fruit tree related to the plum and originating from Central Asia. The apricot has been cultivated in China for about 4,000 years, and 2,000 years ago it came via Armenia and Greece to Italy. The fruit is a stone-fruit. When ripe, it has a fine flavour; but if it is to be transported, it is picked while un-ripe and will then never get its fine aroma. Dried and preserved apricots are an important product. The seeds contain 40% of oil and are used as a substitute for almonds. Apricots are mainly grown in sub-tropical, dry districts, e.g. in California, the countries around the Mediterranean, and the oases of Western and Central Asia. In England the climate is too cold and damp for their cultivation.

85 Peach *Prunus persica*

A tree reaching 26 ft in height, a near relative of the almond, with its home in East Asia. It was cultivated in China as early as 4,000 years ago, and from there it came to Persia, Asia Minor, Italy and the rest of Europe. The peach is a stone-fruit, and the seed contains 40-50% of oil with a taste of bitter almonds; it is used as a substitute for almond oil. It is grown in the same areas as the apricot and in the warmer and drier parts of England. Dried and canned peaches are an important export article, particularly from California and South Africa.

86 Black Mulberry *Morus nigra*

A tree of average height, originating from the forests south of Caucasus and the Caspian Sea, and known as a cultivated plant in the Mediterranean countries in ancient times. A mulberry is a compound fruit resembling a blackberry in which the flower-leaves or calyx of each female flower swell and become juicy. The half-ripe fruit is red, the ripe fruit is almost black and nicely flavoured. Mulberries are grown in Central and South Europe. In this country they are grown only occasional-ly, in the south. The white mulberry, *Morus alba,* has white, rounded fruits; it comes from East Asia and Himalaya, and is very important for the silk industry, as the silk worms feed on its leaves.

87 Fig *Ficus carica*

A small tree with milky sap in all its green parts; its home is the Near East and the Mediterranean countries. The fig has been grown since ancient times. A fig is a complete fruit consisting of small seeds containing fruits with thin skins inside a sort of hollow globular or pear-shaped receptacle which swells and becomes sweet and juicy when ripe. The cultivated fig *Ficus* varieties contains female flowers only. The *Caprificus* variety was formerly consider-ed to be the wild species, but is now regarded as carrying the male flowers of the cultivated fig. The *Caprificus* variety also contains female flowers (gall flowers), in which the gall fly is hatched. These insects are essential for the pollination of cultivated figs and for seed setting. Consequently branches of *Caprificus* are grafted on the cultivated fig trees in order to ensure pollination. However, it is possible, except in the case of the Turkish or Smyrna figs, to obtain large edible figs without pollination.

Fresh figs have a sweet, and to some people attractive, taste. Dried figs are an important article of commerce. The finest figs come from Asia Minor (Smyrna figs), but good varieties are also supplied by Spain, Italy, Southern France, Israel and North Africa. Greece contributes ring figs, strung on rush string. Italy has the largest production.

88   Sweet Orange *Citrus sinensis*

A 16-40 ft high, evergreen tree, originating from the sub-tropical summer rain areas in South-east Asia. It has been grown in China for thousands of years. The fruit is a many-celled berry with a thick outer rind and a spongy inner rind; the segments contain seeds surrounded by juicy pulp. Some kinds, known as blood oranges, have red juice, others have no pips, e.g. Jaffa and Malta oranges. Neroli and Portugal oils, used in perfumery, can be extracted from the fragrant flowers. Oranges are mainly grown in sub-tropical winter rain areas all over the world, and are artificially watered. Spain, Italy, Israel and the U.S.A. (California and Florida) are the chief exporters. The world production of oranges amounts to 7-10 million tons annually. Orange juice and orange marmalade are also important articles of world trade.

89   Mandarin *Citrus nobilis deliciosa*

A bush or low tree, coming from China. Like the orange, its fruit is a many-celled berry with a thick outer rind,

spongy inner rind, and the seeds are surrounded by a juicy pulp. The mandarin rind comes off more easily than that of the orange, and the fruit pulp has a perfumed taste. The mandarin does not last so well as the orange, and is consequently not marketed to any appreciable extent. Only in 1830 was it first cultivated in the Mediterranean countries, and later still in America. Italy and North Africa are the largest exporters.

90   Grapefruit *Citrus paradisi*

A much-grown citrus variety developed in America from a form of shaddock, a large citrus species from the East Indies. The grapefruit is a many-celled berry like the orange, but it is much larger, weighing about 1 lb. Its pulp is bitter-sweet and acid and it is usually eaten with sugar. Italy and North America have the largest exports; both the fresh fruit and tinned juice are exported from Israel.

91   Lemon *Citrus limonia*

An evergreen tree, coming from Indo-China which made its appearance in the Mediterranean countries 1,000 years ago. It is very similar to the orange, but flowers all the year round so that lemons can be harvested at all times. The lemon is also a many-celled fruit with a thick outer rind and a spongy inner rind. In the cells the seeds are surrounded by a juicy pulp. Lemons contain large amounts of citric acid, and are usually not eaten raw, but are pressed and the juice extracted, which is very rich in vitamin C, is used for drinks and as a flavouring. Lemons are also used for the manufacture of citric acid and lemon essence. They are extensively grown in Spain, Italy and California, where they are picked and despatched while unripe.

92 Carob or Locust Bean
Ceratonia siliqua

A small evergreen tree from the eastern countries around the Mediterranean. The edible part is the pods, 4-18 in. long. These are the true locust beans. They are flat with a hard shell, sweet pulp (sugar content 50%), and large, hard seeds (carats) which in ancient times were used as weights by jewellers and goldsmiths. Carat (35 grains) is still the unit of weight for diamonds and the measure for purity of gold in alloys, pure gold being 24 carat. The locust bean is of a certain importance as food for the people of the Mediterranean countries. It is also used as fodder.

93 Pomegranate *Punica granatum*

An evergreen, originating from South-west Asia, and a very old cultivated plant. The fruit is of the size of an apple, many-celled and with firm, leathery skin; it has numerous seeds with a fleshy outer coat which constitute the edible part of the fruit. The taste is acid-sweet. It is grown in areas in Western Asia and the Mediterranean countries where the climate is too dry for citrus varieties to do well. The bark and the skin of the fruit contain an effective agent against tape-worm; in India the bark is used for treating dysentery.

94 Date Palm *Phoenix dactylifera*

A 65 ft high, heat-loving palm, originating from the African-Asiatic belt of deserts, where it forms the essential basis for human existence. It has been grown for 4-5,000 years. Date palms are indispensable in the oases, where they afford shelter from the sand-storms, and shade for the gardens. The date palm must have dry heat up above and moisture down below; its roots penetrate very far, all the way down to the subsoil water. Hundreds more female plants than male are grown, bearing fruit from their 6th to their 100th year. The fruits (dates) are single-seeded, berry-like stone-fruits, containing 70% of carbohydrates and nicest to eat when dried. The seeds are used as camel-fodder or, roasted, as a coffee substitute. Dates are a very important food for the nomads, who also eat the fresh shoots of the palm-top as palm-cabbage. Ropes can be braided from the fibres of the petioles and the trunks can be used for timber. The best dates come from Iraq. Dessert dates, packed in boxes with their stems, come from Tunisia and Morocco. The only country in Europe where dates can be grown profitably is Spain, where they were introduced by the Moors.

95 Mango *Mangifera indica*

A 65 ft high tree with a wide crown, one of the most important fruit trees of the tropics. It originated from tropical East Asia and is unknown in a wild state. The fruit, the mango, is a stone-fruit with a peculiar, frayed, flat stone. The mango varies considerably in shape, colour, taste and size (2-12 in.). The largest weighs 5 lb. The fruit pulp contains 12-19% sugar; its taste resembles that of turpentine, and it is eaten fresh, or preserved as mango chutney.

96 Persimmon *Diospyros kaki*

A small tree related to ebony, with its home in East Asia. Its fruit, the persimmon, is a tomato-like berry with a taste resembling that of apricot; it should preferably be eaten over-ripe when the pulp is liquid. It is extensively grown in China, Japan, California and Southern Europe, but it cannot tolerate a tropical climate.

97 Common Papaw *Carica papaya*

10-32 ft high, palm-like tree with milky

sap; its home is in Mexico. It flowers and bears fruit all the year round, but often dies as early as in its fourth year. The fruit, the papaw, is a large, melon-like berry with green, yellowish, or reddish skin; the fruit pulp is yellowish, sweet and spicy. It has numerous seeds with an acrid taste in the wall of a large, central cavity in the fruit. The papaw exists in many varieties and sizes, its weight varies from 1-24 lb, and it is often eaten raw with sugar, or boiled. Excessive eating of the papaw may result in blood suffusion in the skin owing to the large content of papain. This is used in medicine as a horn solvent, and has a further use in making meat tender. The papaw is grown all over the tropics.

98   Avocado Pear *Persea gratissima*

A tree of the bay family, originating from tropical America; it was grown long before the discovery of America, i.e. among the Aztecs. Its fruit, the avocado pear, aguacate, or alligator pear, is a large, savoury berry. It has a large, bitter seed. The consistency of the avocado pear is like butter, and its taste resembles that of fine nuts. It is usually eaten either with pepper and salt, or with lime juice, sugar, etc. Avocado pears vary considerably in shape (apple-, pear-, or cucumber-shaped), in colour (green, purple, and black), and in size (¼ lb-4 lb).

99   Banana *Musa sapientum*

Up to 32 ft high, tree-like plant, with its home in the tropical rain forests of South east Asia. Ancient plant cultivated in India; only in the 16th century did it reach America via the Canary Islands. The false trunk is formed by the leaf-sheaths. The fruit (banana) is a seedless berry with a tough, leathery skin. The fruits are clustered in a 'hand' of bananas, with 14-20 bananas in each cluster; these, in turn, grown in clusters of 6-9 'hands' in a cone of large, vividly coloured leaves (bracts) which, however, fall off long before the fruit is ripe. The banana plant needs plenty of rain all the year, and does best in tropical rain forests. The banana is the most important food for many tropical peoples, e.g. the Bantus in the Congo, and is one of the most important articles of the international fruit trade. Each surface shoot will produce only one cluster of fruits, weighing approximately 55 lb, which is harvested while it is green. The main area of banana-growing is in Central America along the Caribbean coast. Tropical South America, tropical Mexico and Jamaica are other important export countries. Outside America, only the Canary Islands and the Guinea coast are important export countries. The U.S.A. absorbs two-thirds of the world production.

100   Pineapple *Ananas sativus*

A herbaceous tropical plant, with its home in Brazil; it was grown all over tropical America before the discovery of that continent. The plant has a certain resemblance to the sisal plant (agave); it has a leaf rosette with a short-stemmed, egg-shaped spike of small flowers, covered by fairly large scales. On ripening, not only do the individual fruits become juicy, but also the entire spike with stem and scales, which combine to form the large collective pineapple, weighing anything from 1-33 lb. After fruiting the whole plant dies. It can only be propagated vegetatively by the shoot on top of the fruit or by suckers as there are no seeds. Pineapple is now grown in nearly all tropical countries, and is exported mainly tinned; its importance on the world market is increasing. 75 % of the world production comes from Hawaii;

but the best quality is produced on the Sunda Islands. The leaves may be used for yarn and fine materials (pineapple cambric or ananas cambric), but the plants cannot also be used for fruiting at the same time.

101 Musk Melon
Cucumis melo var. reticulatus

A climbing annual, closely related to the cucumber, and probably originating from Indonesia. It has been grown since ancient times. Its fruit is a berry with yellow or orange flesh. In Britain it is grown only in greenhouses or in frames.

102 Water Melon *Citrullus vulgaris*

A giant relation of the melon, growing in a wild state in South Africa, which was cultivated in the Mediterranean countries before the Christian era. The water melon has yellow or red flesh, and its taste is sweet and slightly flat; the seeds contain oil and are nourishing. The wild varieties form a mighty vegetation in the Kalahari Desert, providing the population with the necessary liquid to survive during the dry season. It is widely grown in Hungary, Southern Russia and the U.S.A., and all over the world in the tropics and sub-tropics.

103 Red Currant *Ribes rubrum*

A 3-7 ft high bush. The cultivated varities originated from several different forms of currant, all closely related, from Northern and Central Europe. They were first grown at the end of the Middle Ages. The currant fruits are round berries in a cluster; they contain sugar, citric acid, vitamins A and C, and are much used for juice, jelly and jam. The *Ribes* varieties are mainly propagated by layering.

104 White Currant
Ribes rubrum var. leucocarpum

A variety of currant with white berries, less sour than red currants.

105 Black Currant *Ribes nigrum*

The wild-growing primitive form is common in forest marshes in Northern and Central Europe and eastwards to China. This plant resembles the red and white currants, but the green parts are covered with small, yellow glands, producing the characteristic smell. The berries are black, and contain large amounts of vitamins A and C, and are used for jam, jellies and liqueur.

106 Gooseberry *Ribes grossularia*

A thorny bush, growing in its wild state over nearly the whole of Europe. It has been grown in England since the time of Henry VIII. The fruit is a berry of widely varying shape, colour and hairiness; it contains sugar, citric acid and vitamins A and C. Gooseberry bushes are often subject to serious attacks by mildew, known as American gooseberry mildew as it came from America to Europe fifty years ago.

107 Garden Raspberry *Rubus idaeus*

A bush growing in its wild state in several varieties nearly all over the temperate areas of Europe and Asia. It was cultivated from the end of the Middle Ages and is known in numerous varieties with larger fruit; however, their aroma is not as fine as that of the wild varieties. A raspberry is a composite fruit consisting of small seed containing fruits (drupelets) which may be red or yellow. The part of the flower carrying the fruit remains as a white core in the raspberry when it is ripe.

108 Blackberry *Rubus fruticosus*

A common name for a large number of

semi-bushes with long rambling growths, growing wild in leafy woods, wood edges and along hedges all over the temperate zone of the Northern Hemisphere. The fruit is composite consisting of small seed containing fruits (drupelets) similar to the raspberry; when ripe they are a shiny black, juicy and sweet. Blackberries were used centuries ago but have only been cultivated during the last hundred years.

109   Dewberry *Rubus caesius*

This grows in its wild state all over forest and coastal regions. It has a close resemblance to the blackberry. It is a composite fruit with few, fairly large, matt blue little fruit (drupelets); its taste is rather flat.

110   Arctic Raspberry *Rubus arcticus*

6-10 in. high, closely related to the Cloudberry, growing round the Polar Circle in the Northern Hemisphere. The fruit is a composite consisting of small seed-containing fruits, which are dark-red and delicious to eat.

111   Cloudberry *Rubus chamaemorus*

A 4-10 in. high plant with male and female flowers, related to blackberries and raspberries. Grows wild in North America, Northern Asia and in Scandinavia, on high moors. The cloudberry is a composite fruit consisting of small, juicy drupelets rich in vitamin C, and with a fresh, acid taste. Cloudberries are hard and red when unripe, but when they ripen they become soft and yellow.

112   Wild Strawberry *Fragaria vesca*

A wild-growing strawberry, common across Europe in woods and thickets. The fruit is a false fruit, consisting of many small seeds on a swollen, juicy flower base. Wild strawberries have a fine aroma and flavour and have been

grown since ancient times, but have now been almost supplanted by varieties bearing larger fruit, arising by hybridisation of certain American species.

113   Alpine Strawberry
Fragaria vesca var. semperflorens

A cultivated variety of wild strawberry with larger fruit which have, however, kept their fine aroma and flavour. Alpine strawberries are perpetual and generally bear flowers and fruit from May to the autumn. They may or may not have runners; those without may be propagated by division.

114   Garden Strawberry
Fragaria x ananassa

This is the cultivated strawberry created around 1800 by improving and crossing several American varieties, which produced the varieties with large berries called garden strawberries. The crop from strawberry cultivation depends to a large extent on the weather, both in the flowering and ripening seasons. 'Royal Sovereign' is a fine table berry, also 'Auchincruive Talisman' and the 'Cambridge' varieties.

115   Sloe *Prunus spinosa*

A thorny bush, related to the plum; it grows particularly along hedges and forest edges. The fruit is a stone-fruit with an astringent taste, due to a large content of tannin. When the fruits have been exposed to frost, the taste becomes less sharp, and they may be used for jam and sloe gin.

116   Dog Rose *Rosa canina*

A wild-growing rose. Its fruit, the hip, is a false fruit, with many small, hairy nuts in a coloured, jar-shaped flower base. Wild hips of many different rose varieties are gathered in many areas in Europe, and sometimes used for soup. Vitamin preparations are made from them, as they are very rich in vitamins A and C. The imported *Rosa rugosa,* with large, round hips, is also used for this purpose. It is used as a stock for garden roses.

117 Bilberry, Whortleberry, or Blaeberry *Vaccinium myrtillus*

An 8-20 in. high, deciduous dwarf bush, growing wild in Europe, Asia and North America, common on moorlands and thickets in peaty soil. The dewy, blue berries, rich in vitamin C, have red or purplish juice; they are eaten fresh, or are used for pies. The juice turns blue when exposed to an alkali.

118 Cowberry *Vaccinium vitis-idaea*

An 8 in. high, evergreen dwarf bush, growing wild on moors, high moors and pine forests in the northern areas of the whole Northern Hemisphere. The fresh berries have a bitter taste, as they contain benzoic acid; but they are used for jam in countries where they are plentiful enough, such as Sweden.

119 Cranberry *Vaccinium oxycoccus*

An evergreen, creeping, threadlike bush, growing wild on high moors around the Polar Circle and grown commercially in North America. The berries are larger and clearer than cowberries, and are eaten fresh, with sugar or cooked. They may also be bottled. Some berries imported as cow-berries are actually cranberries.

FODDER PLANTS

120   Timothy Grass or Cat's-tail
Phleum pratense

A variety of grass, in its wild state growing in Europe and neighbouring areas in Asia and Africa. It is easily recognized by its stiff tuft of flowers. About 1720, the Danish emigrant Timothy Hanson began to cultivate it in America, whence it was transferred to Europe. It is widespread in the British Isles and used for grazing and hay.

121   Meadow, or Common Fox-tail Grass *Alopecurus pratensis*

A variety of grass resembling the foregoing, but with a softer tuft of flowers, tapering towards the top. It is grown for fodder and does best in water meadows and other damp soil. The meadow fox-tail develops early and affords good pasture after cutting.

122   Italian Rye-grass
Lolium multiflorum

A valuable fodder grass, originating from Southern Europe, introduced into Britain about 1830 and much grown at present. Common rye-grass, *Lolium perenne* is also grown for fodder, and is often used for lawn grass.

123   Cock's-foot
Dactylis glomerata

A tall, perennial grass, easily recognised by its crowded flower head. It has rapid growth and is a fairly good fodder grass, sometimes replacing timothy grass. Much of the seed comes from Denmark. Cock's-foot is unsuitable for permanent pasture or lawns because of its tufted growth.

124   Common Clover
Trifolium pratense

A perennial leguminous plant, growing wild all over Europe, Western Asia and North-West Africa. It was first cultivated in the 12th and 13th centuries in Italy and Spain, whence it spread to the Netherlands and Germany, and later to Great Britain. This and the following leguminous plants (Nos. 125-130) are particularly valuable, as they can obtain nitrogen for themselves in soil poor in nitrogen, their roots being equipped with bacterial nodules which are able to absorb the air's nitrogen, and this is then used by the plant. Pollination is effected by bumble-bees (humble) with long proboscis; these are attracted in large numbers by the colour and fragrance of the clover. Common clover is one of the most important fodder plants; there are numerous varieties, both early and late. It will yield a large crop only the first year; but it may be harvested twice yearly.

125   Alsike Clover or Shamrock
Trifolium hybridum

A clover variety from Europe, North Africa and Western Asia, described in 1742 by *Linné*, who recommended it for cultivation. Aslike clover resembles the common clover and is excellent for pastures when ground is too wet for common or red clover.

126   White Clover or Dutch Clover
Trifolium repens

A low, creeping clover variety, growing wild in Northern Europe and Asia but improved by cultivation. It is unsuitable for cutting, but good as pasture, as it will tolerate trampling by cattle.

127   Common Vetch or Tare
Vicia sativa

An annual leguminous plant, cultivated

since ancient times and often sown in a mixture with oats for green fodder. The common vetch is harvested while it is blossoming, before the pods have developed.

128 Lucerne or Alfalfa
Medicago sativa

A 20-60 in. tall, perennial leguminous plant, originating from the western temperate areas of Asia, i.e. Persia. It has been grown since ancient times, and came to Greece during the Persian Wars, and thence to Italy. The Moors who brought it to Spain, called it alfalfa. In the middle of the 18th century it arrived in Scandinavia. Lucerne will tolerate dry summers, as its roots penetrate up to 6-7 ft into the earth. It makes great demands on the soil and will not tolerate grazing but it may be cut 2-3 times a year. It is widely grown in the U.S.A. It is used for green fodder and hay, and for the manufacture of lucerne meal, or flour, which contains vitamin K. Lucerne is also used for feeding small domestic animals.

129 Yellow Lupin *Lupinus luteus*

A 15-35 in. high, annual fodder plant, coming from Southern Europe. Like all the foregoing fodder plants of the leguminous family, it is valuable because it can provide its own nitrogen on soil poor in nitrogen. Its roots are equipped with bacterial nodules able to use the nitrogen in the air. This nitrogen is then assimilated by the plant with the help of the bacteria.
The lupin may be used as a fertilizer by ploughing the entire plant into the soil at the end of August. The lupin contains lupinine, which is bitter-tasting and poisonous, but when made into silage this disappears, so that the plant may be used for fodder. By selection of the varieties containing the smallest amounts of poison over several

generations, almost lupinine-free varieties have been obtained. These are called sweet lupins, which may be used for fodder directly, and the oily seeds may be used for the manufacture of edible oil. Lupins will grow best on poor soil without much lime, and tolerate cold climates well.

130 Sainfoin *Onobrychis viciifolia*

A perennial fodder plant, with its home in the Mediterranean countries. It is used in the same way as lucerne, but has been superseded as the latter has become more extensively cultivated.

131 Fodder Beet
Beta vulgaris ssp. rapa

A biennial beet, partly a primitive form of the sugar beet, but with a much thicker first-year root jutting out from the soil. If it is to be used for seed production, the plants must be taken out after the first year, and planted out again in April. The fodder beet is of great importance as cattle fodder in most of Europe. There are several varieties.

132 Turnip
Brassica rapa var. rapifera

A biennial cultivated form of charlock; it was cultivated in olden times, and known by the Romans. It is less exacting than the fodder beet, but less nutritious, and is being supplanted to an increasing extent by the swede. A field of turnips may be recognized from afar by its grass-green, not dewy-blue, leaf rosettes, as distinct from the swedes. There are varieties with white flesh as well as with yellow; the one shown here is a 'White Tankard'.

HERBS

133 Parsley
Petroselinum hortense or crispum

A biennial umbellifer, originating from Spain and Greece, where it was grown in ancient times. All parts of the plant are rich in essential oils, making it suitable for seasoning; it contains much vitamin C and some iron. The leaves are the part most generally used, for this purpose curled parsley has come to be grown almost exclusively here in Great Britain. The 8 in. thick root of the Hamburg or turnip-rooted parsley, *var. tuberosum,* is sometimes used in soups.

134 Dill *Anethum graveolens*

An annual umbelliferous plant, originating from the Mediterranean countries. It is little used in this country, but in Denmark the umbel with its almost ripe fruits is used in flavouring cucumber and gherkins. In Sweden, the plumy leaves and young shoots are used to a very wide extent as seasoning for many dishes, particularly herring and crayfish.

135 Caraway *Carum carvi*

A biennial umbelliferous plant growing widely from Europe to North India. The aromatic fragrance of the seeds is due to their content of caraway oil, 1-4%, which is used for flavouring aqua-vitae and a liqueur (kummel). The whole seeds are used in this country for flavouring and abroad in cheese, rye-bread, etc.

136 Fennel *Foeniculum vulgare*

A biennial, umbelliferous plant, which is grown to a certain extent around the Mediterranean, where its leaves are used in fish sauces, for garnishing, and its seeds for flavouring, like caraway.

The fennel, which is rich in anise oil, is also used for medical purposes in mixed liquorice powder.

137 Finnochio or Florence Fennel
Foeniculum vulgare var. dulce

A cultivated variety of No. 136, with its home in Southern Europe, especially Malta. The seeds have a milder taste than of the fennel. The first year's leaf rosette is used, the base of which is swollen and forms a sort of tuber, which cooked in stock is eaten as a vegetable; flavour like sweet celery.

138 Aniseed *Pimpinella anisum*

An annual, umbelliferous herb originating from the countries to the east of the Mediterranean. The two halves of the seed can be separated only with difficulty, and have soft hairs, which have generally been worn off by the time they are sold. The characteristic taste is due to anise oil which, among other things, is used in anisette liqueur and in boiled sweets (aniseed balls). The dried, pulverized seeds are used in medicine, in cough mixtures and weak opium drops. Aniseed is grown in Southern Europe, Russia, India, Japan and South America. Several other plants also supply anise oil.

139 Chervil *Anthriscus cerefolium*

An annual umbelliferous herb, growing wild in Central Europe and Southern Europe. The small, fine, fresh leaves have a peculiar smell and taste due to an essential oil; they are used for salads, garnishing and seasoning.

INDEX OF ENGLISH NAMES

References are to page numbers

INDEX OF LATIN NAMES

References are to page numbers